# GLP Essentials

## A Concise Guide to Good Laboratory Practices

Second Edition

Milton A. Anderson

informa
healthcare

New York London

Informa Healthcare USA, Inc.
52 Vanderbilt Avenue
New York, NY 10017


Printed in the United States of America on acid-free paper
10 9 8 7 6 5

International Standard Book Number-10: 1-5749-1138-4 (Hardcover)
International Standard Book Number-13: 978-1-5749-1138-1 (Hardcover)



**Library of Congress Cataloging-in-Publication Data**

GLP essentials: a concise guide to good laboratory practice--2nd, ed. Milton A. Anderson
p. ; cm.
Includes bibliographical references and index.
ISBN-13: 978-1-5749-1138-1 (alk. paper), ISBN-10: 1-5749-1138-4 (alk. paper)
1. Laboratories--Technique--Handbooks, manuals, etc. 2. Laboratories--Standards--Handbooks, manuals, etc.
[DNLM: 1. Laboratory techniques and Procedures--standards--Laboratory Manuals. QY 25 A548g 2002] I. Title.

Q183.A1 A53 2002
542'.1'0685--dc21 2002014193

**Visit the Informa Web site at**
**www.informa.com**

**and the Informa Healthcare Web site at**
**www.informahealthcare.com**

# Contents

# Preface

This booklet is an introductory presentation of good laboratory practices. It is intended as a primer for those who are involved or associated with understanding the needs for improving the integrity of scientific documentation. It provides a historical background of Good Laboratory Practice (GLP) development worldwide and the rationale for establishing these requirements in a rapidly expanding scientific research and regulatory environment. Additionally, it identifies and describes the basic required elements of managing a scientific study–its planning, performance, reporting, and monitoring. After reading this publication, the reader will be aware of the origin and implementation of GLP, and will have a greater appreciation of the benefits produced by its implementation in industry, academia, and regulatory agencies.

It is anticipated that this Guide will serve as an incentive for you to continue your interest in good laboratory practice. Besides the literature references,* this edition also includes Internet Web-site identities that will provide broader and more expeditious access to historical as well as current publications. It is hoped that this introduction to GLP will provide for a more thorough understanding of good laboratory practices and, perhaps, serve as a building block in your career development.

*See Anderson 2000, Hirsch 1989, and Carson and Dent 1994 for a more thorough discussion of GLP.

# Chapter 1
# Introduction

Good laboratory practice today can be applied to very broad fields of science. Whether you are involved in chemical, pharmaceutical, or ecotoxicological branches of science, knowledge of good laboratory practice is essential in establishing and maintaining the integrity and high quality standards of today's scientific developments. It is now applicable in laboratories of academia, industry, and governments throughout the world. Good laboratory practice has been defined as the management of scientific studies through the planning, performance, and reporting phases, with ample support through compliance monitoring.

## Early Recognition of Good Laboratory Practice

Good laboratory practice is a relatively new term, but it has been a subject appearing in numerous pharmaceutical and ecotoxicological articles, primarily during the past decade. Elements of it, however, have been around for many years. Science teachers usually applied it as a set of rules for proper note keeping. These rules, usually about ten, were listed at the front of each student's laboratory notebook. Included would have been directions such as:

- All entries must be made in ink.
- Erasures are not permitted.

- Entries should be dated, signed, and witnessed.

Routinely, these notebooks were evaluated and graded. If you have remained in research, working in a government, industry, or university laboratory, these rules continue to apply, routinely serving as criteria comprising part of your performance review.

Proper scientific documentation also remains very important in today's business environment, motivated by patent requirements for protecting your ideas, and those of your company or institution. There are now, however, other equally important factors that are accelerating the need for an expansion of these elements, as reflected in the concept of good laboratory practices worldwide and described in this publication.

### Government Involvement

In the early 1970s, several governmental agencies, including the US Food and Drug Administration (FDA) and the US Environmental Protection Agency (EPA), became aware of the need to develop formally a more elaborate set of good laboratory practices that go beyond just appropriate documentation to become involved in the performance and confirmation of the integrity of the scientific studies themselves. These agencies, whose charter has been to protect public health, recognized that product safety can only be assured by confirming that supportive research studies are conducted in accord with acceptable practices. Early investigations by these agencies had concluded that some regulatory submissions were tainted with dishonest data that if undetected, would have led to approval of an inherently unsafe product, with potentially very serious, long-term impacts on public health.

### Research Study Integrity

Confirmation of appropriate study conduct and documentation was needed, and the regulatory decision process had to be based on data of the utmost credibility. After extensive investigative evaluations of academia and industry, these regulatory agencies developed and implemented the GLPs as guidelines, standards, or regulations. These now serve as the means of providing scientific reports of high quality and integrity to regulatory agencies, allowing them to confidently provide decisions regarding a product's premarket safety and efficacy, and, therefore, to protect public health worldwide.

### The Good Laboratory Practice Concept

Among the first countries to formally address good laboratory practice was New Zealand, with the 1972 promulgation of the New Zealand Testing Laboratory Registration Act. This act defined the testing laboratory to include "the equipment, facilities, staff, records, procedures, and places used in testing." Additionally, it defined *test* as "determining in whole or in part, the composition or physical properties of a substance or product, calibrating a piece of equipment or determining the ability of any substance, product, or piece of equipment to satisfy particular requirements." The act also established a Testing Laboratory Registration Council, with functions and powers "to promote the development and maintenance of good laboratory practice in testing."

In March 1973, Denmark approved legislation that addressed laboratory practices. The "Law About the State's Technical Trial Board" defined the duties of the Board as advancing, promoting, coordinating, and keeping control of authorized technical experiments for the purpose of bringing about safety and quality control (Danish National Testing Board Act 1972). The research addressed includes that related to a material's physical, chemical, structural, and quality characteristics, and how the material performs under various environmental conditions. The Board, the law explained, monitors the progress of authorized research activities, both public and private, including the auditing of documentation of issued research results. Approved research laboratories must satisfy certain criteria relating to the knowledge and competence of the personnel, the capacity of the organization, and the suitability of research equipment.

Prior to this legislation, there were many private and public laboratories worldwide that had established specific means to confirm the credibility, integrity, and quality of research. These voluntary policies were intensely enforced in some, but not all, organizations. In many countries, government was becoming more involved in both the awarding of financial grant programs for research and the requirement of government approval prior to marketing materials and products that affect the well-being and safety of many people. These government bureaus recognized that any decision to approve or disapprove the marketing of a product was only as good as the reliability of the research and data submitted to them. The significance of the New Zealand and Danish approaches was that these countries for the first time had legislated the beginning elements for prescribing laboratory research and reliability, through the concept of good laboratory practices.

The need for major improvements in the authenticity and reliability of laboratory studies in the United States was recognized in the mid-1970s. In 1975, in an existing "for cause" compliance program, the FDA noticed, during the inspections of some firms, deficiencies that seriously impugned the integrity of reported data. As a consequence of these findings and subsequent congressional action, the Good Laboratory Practice for Nonclinical Laboratory Studies were implemented in June 1979 (US FDA 1979).

# Chapter 2
# GLPs in the United States

## Historical Perspective

The need for standard GLP criteria was well established in the United States in the mid-1970s. Although it has been said that these regulations were arbitrarily promulgated to extend the agency's enforcement base, most now concur that these regulations have been purposeful and beneficial. History proves an opportunity to evaluate the conditions reported by the surveillance teams and to judge whether the GLP regulations were necessary. A review of these findings provides an important historical perspective and an instructive first step in recognizing undesirable conditions that may still exist in some laboratories not working to GLP standards. Such a review may help in initiating action to prevent the continuation of or future reemergence of these conditions.

Some of the laboratory conditions existing during the mid-1970s were not necessarily the norm, but were found by the FDA and the EPA during their review of industry operating practices. The review was part of an extensive surveillance of the pharmaceutical and chemical industry. These findings were initially reported in testimony to the United States Senate Subcommittee on Health in 1975 and 1976 and were subsequently publicized in newspapers and scientific journals. The most prominent evidence supporting the need for regulations was the flagrant research discrepancies noted by the FDA and the EPA at a contract research laboratory. The laboratory, Industrial Biotest Corporation (IBT) of Northbrook, Illinois, was a leader in the contract testing field, performing over twenty thousand studies to support the safety and efficacy data generated for hundreds of drugs and pesticides. The alleged irregularities in the data were numerous. They included the falsification of laboratory work, replacement of animals that died under test with fresh animals without documenting the substitution, fabricating test results, and excluding test results if the results were not considered favorable by top company officials. Three company officials were found guilty of defrauding the government by falsifying drug and food additive research data. Their appeal to the Supreme Court was denied in 1986, and all were given lengthy jail sentences.

### Evaluation of Research Integrity

Governmental surveys of nonclinical studies performed at various types of institutions revealed vast differences in the means of documenting study reliability. At the time of the government compliance program survey, the institutions audited were not aware of the evaluation criteria. The agency had contended that the study data submitted in support of a regulated product were based on appropriate experimental procedures by a testing facility. Industry, contract laboratories, and academia were subsequently audited in the absence of published regulations, contributing widely to "compliance" disparities.

It is important to be aware of these governmental surveillance observations and conclusions, particularly where regulations or guidelines, such as the GLP regulations, have not been implemented in a company performing laboratory studies. These observations and conclusions provide the rationale for the need and subsequent design and development of the GLP regulations or guidelines. Some of the FDA's observations are described here for your information.

### *Surveillance Program Observations*

- Original autopsy records for certain studies were either unavailable or apparently transcribed to new records several years after the autopsies.
- Pathology reports submitted to the agency were inconsistent with original autopsy records.
- Microscopic examinations of tissue slides were conducted by more than one pathologist, each of whom came to different conclusions; yet only the conclusions favorable to the drug were submitted to the agency.
- Records of laboratory observations were neither dated nor signed.
- Employees were unable to account for discrepancies between raw data and final reports submitted to the agency.
- Animals were observed and recorded as normal for a variety of factors, including appearance, appetite, and thirst, when in fact the animals were dead.
- Drugs under study were administered to animals in a manner that made it impossible to determine how much, if any, of the required dosage was actually ingested by the animal.

- At one firm, the reproduction and teratology studies were conducted by laboratory personnel overseen by a senior scientist who did not have the proper qualifications or background to be conducting and supervising these critical studies.
- In another case, necropsies were being performed by people without the proper training, as was recognized by a senior scientist who reviewed the work.
- Treatment and control animals were not properly identified.
- Animals were fixed in toto and not necropsied for several months.
- A laboratory was sprayed and fogged with pesticides while animals were in the laboratory.
- In one study, the FDA was told that animal tissue had been examined histopathologically, when a review of the contract laboratory's original records indicated that these tissue samples were never even collected.
- Significant discrepancies were found between gross observations on pathology sheets and with the individual pathology summaries submitted to the agency.
- One firm submitted a study utilizing the wrong data and the wrong animal identification numbers, which were easily discovered by the agency, indicating that management did not check the data used.[1]

[1]Federal Register, Vol. 41, No. 225, Friday, November 19, 1976, Department of Health, Education, and Welfare, Food and Drug Administration, Nonclinical Laboratories Studies, Proposed Regulations for Good Laboratory Practice, pp. 51207–51208.

These perhaps represent many of the extremes of the observations that had surfaced during the surveillance of various institutions. From this type of evidence, it is apparent that the agency had the right and the responsibility to be concerned about the submissions they were required to review and provide judgment on regarding the approvability of the product.

These observations, along with others, provided the FDA and the EPA important information for developing conclusions that contributed to the development of GLPs in the United States and standards throughout the world.

### *FDA's Conclusions*

- Experiments were poorly conceived, carelessly executed, or inaccurately analyzed or reported.
- Technical personnel were unaware of the importance of protocol adherence, accurate observations, accurate administration of the test substance, and accurate record-keeping and record transcription.
- Management did not assure critical review of data or proper supervision of personnel.
- Assurance could not be given for the scientific qualifications and adequate training of personnel involved in the research study.
- There was a disregard for the need to observe proper laboratory, animal care, and data management procedures.
- Sponsors failed to monitor adequately the studies performed by contract testing laboratories.
- Firms failed to verify systematically the accuracy and

completeness of scientific data in reports of nonclinical laboratory studies before submission to the FDA.[2]

These observations, together with other criteria, were used in the development of the November 1976 proposed GLP regulations. The FDA, however, also initiated another surveillance program directed toward uninspected testing facilities to determine if there were "proposal" problems they had not anticipated.

### *Results of Evaluations*

Those in the healthcare field were exposed early to these governmental compliance surveys (Department of Health 1975). This survey, conducted for approximately three months, consisted of 39 nonclinical laboratories, 23 sponsor laboratories, 11 contract laboratories, and 5 university laboratories. These audits were used to acquire information for fine-tuning the final governmental regulations. The Department of Health and Human Services reported in *Analysis of Sponsor Laboratories for Compliance with the GLP Regulations* the results of this program, indicating some disparity in compliance between industry and academia: 69 percent of the requirements were met by sponsor laboratories, 56 percent by contract laboratories, and 46 percent by university laboratories.

Primary noncompliance with the proposed regulations occurred in facilities failing to have a quality assurance unit (QAU), failing to test each batch of test article-carrier mix-

---

[2]See note 1 above.

ture, and lacking standard operating procedures (SOPs). The negative results regarding the QAU should not be surprising, because this was a relatively new concept and in many operations the writing of formalized SOPs was a major task not yet completed.

More recent reports from the FDA indicate that for a six-year period, 1981 to 1987, 60 percent of the 444 laboratories inspected were issued FDA 483s, the agency's official non-compliance document of violative observations. It would appear that this has justified the need for continuing surveillance by the FDA.

### Implementation in the US

As a result of the FDA Bioresearch Monitoring Program, (including the pilot inspection program), published commentary, and public hearings, the US FDA published the GLP regulations, which became effective on June 20, 1979. Since their publication, these regulations have remained relatively constant, but modifications have been implemented, as reflected in the latest *Code of Federal Regulations.*

The US EPA has also been intensely involved in developing GLP standards (US EPA 1983, 1984) for products falling under its jurisdiction. These GLP standards apply to non-pharmaceutical chemicals and are reflected in two distinct standards: one for toxic chemical substances (TSCA–Toxic Substances Control Act), implemented in December 1983 and the other for pesticides (FIFRA–Federal Insecticide, Fungicide, and Rodenticide Act), effective in May 1984.

# Chapter 3
# GLP Development Internationally

## Worldwide Interest

Because the GLP regulations as conceived in the United States had much broader implications than previous acts, they aroused interest in other countries and among international organizations, such as the World Health Organization (WHO) and the Organisation for Economic Co-operation and Development (OECD). In April 1978, the 22 member countries of the OECD, meeting in Stockholm, arrived at a consensus to assign the responsibility of developing GLPs internationally to the OECD. Beginning in an April 1979 meeting in Washington, D.C., and in subsequent deliberations at meetings in Brussels, Paris, and Berlin, the OECD Council approved the *OECD Principles of Good Laboratory Practice* (OECD 1997) on May 12, 1981. These principles of GLP, although similar to the US FDA GLPs, differ in scope, being applicable to "all non-clinical testing of chemicals," whereas the FDA GLPs apply to "food and color additives, animal food additives, human and animal drugs, medical devices for human use, and biological and electronic products" (21 CFR 58.1). However, the FDA GLPs, along with the two EPA standards (TSCA and FIFRA), essentially represent the full scope of the OECD Principles.

There are other countries that have developed or are evolving comparable regulations or guidelines, including Belgium, Canada, France, Germany, Italy, Japan, Korea, Spain,

Sweden, Switzerland, the Netherlands, and the United Kingdom. Additionally, several of these governments have developed bilateral Memoranda of Understanding (MOUs). These MOUs are supplemental agreements allowing for the mutual acceptance between nations of nonclinical data in support of regulatory reporting requirements for various types of pharmaceutical and chemical products. International harmonization is continuing as a major interest. The scope of this cooperative effort, of course, extends beyond the GLPs; it applies to the common use and application of clinical studies (GCPs–Good Clinical Practices), to the acceptance of Good Manufacturing Practices (GMPs), and to nonclinical studies and GLP inspections.

The Fifth International Conference on Harmonization (ICH-5) was held in San Diego, California in November of 2000. At this conference, a final Common Technical Document was completed and, in May of 2001, it was edited further. This final version was published in the *Federal Register* on October 15, 2001. Harmonization continues as a challenge, and should remain an important part of future symposium agenda. The next conference, ICH-6, will be held in Osaka, Japan in the fall of 2003.

### International GLPs

Many of the GLP regulations throughout the world are similar in content, but the products they represent may differ and they may come under the jurisdiction of different governmental agencies. The following listing of country-specific GLPs, scope, and status provides some insight to the reader regarding the extent of interest and the development of GLPs worldwide.

### *Canada*

Agency: **Health Protection Branch**

Title: Standard for Good Laboratory Practice (GLP) in Non-Clinical Laboratory Studies

Scope: Food and color additives, animal food additives, drugs for use in humans and/or animals, medical devices for human use, biological products, radiation-emitting devices, pesticides, and chemical contaminants

Status: Standards have been proposed and published, but they have not been implemented

### *France*

Agency: **Ministry of Social Affairs and National Solidarity**

Title: Good Laboratory Principles

Scope: Drugs for human and veterinary use

Status: Instruction of May 31, 1983, issued by the Secretariat of State for Health

### *Italy*

Agency: **The Minister of Health**

Title: Principles of Good Laboratory Practices for Chemical Substances

Scope: Drugs for human and veterinary use

Status: Effective August 27, 1986

***Japan*** *(Multiple Agencies)*

Agency: **Ministry of Health and Welfare (MOHW)**

Title: Good Laboratory Practice Standards for Safety Studies on Drugs

Scope: Safety studies on drugs, medical and diagnostic devices, cosmetics, and food additives

Status: Effective April 1, 1983

Agency: **Ministry of Agriculture, Forestry, and Fisheries (MAFF)**

Title: Good Laboratory Practice Standards Applied to Agricultural Chemicals

Scope: Toxicology studies on agricultural chemicals

Status: Effective October 1, 1984

Agency: **Ministry of International Trade and Industry (MITI)**

Title: Good Laboratory Practice Standards Applied to Industrial Chemicals

Scope: New industrial chemicals imported or manufactured

Status: Effective October 1, 1985

Agency: **Ministry of Labor**

Title: GLP Standards to Be Observed by Mutagenicity Testing Institutions

Scope: Occupational safety of industrial chemicals

Status: Effective October 1, 1985

### *OECD*

Agency: **Organisation for Economic Co-operation and Development**

Title: OECD Principles of Good Laboratory Practice

Scope: Safety of chemicals with respect to human health or the environment

Status: Adopted May 12, 1981

### *Sweden*

Agency: **The National Board of Health and Welfare (NBI-IW)**

Title: Good Laboratory Practice for Nonclinical Laboratory Studies

Scope: The assessment of the safety of pharmaceutical specialties

Status: Effective December 17, 1985

### *United Kingdom*

Agency: **Department of Health (DOH) in 1988 replaced the Department of Health and Social Security (DHSS)**

Title The Principles of Good Laboratory Practice

Scope: Health and environmental safety testing of pharmaceuticals, agrochemicals, cosmetics, food additives, and industrial chemicals

Status: Effective April 1983

### ***United States*** *(Multiple Agencies)*

Agency: **Environmental Protection Agency (EPA)**

Title: Good Laboratory Practice Standards (40 CFR 792–TSCA)

Scope: Studies relating to health effects, environmental effects, and chemical fate testing

Status: Effective December 29, 1983

Agency: **Environmental Protection Agency (EPA)**

Title: Good Laboratory Practice Standards (40 CFR 160–FIFRA)

Scope: Pesticides regulations by the EPA

Status: Effective May 2, 1984

Agency: **Food and Drug Administration (FDA)**

Title: Good Laboratory Practice for Nonclinical Laboratory Studies (21 CFR 58)

Scope: Food and color additives, human and animal drugs, medical devices for human use, biological products, and electronic products

Status: Effective June 20, 1979

Other countries are continuing to develop and improve their GLPs, but in the interim, if they do not have a unique document, many are following the standards associated with the OECD GLP principles. Member countries of the OECD now total 29. These include Australia, Austria, Belgium, Canada, the Czech Republic, Denmark, Finland, France, the Federal Republic of Germany, Greece, Hungary, Iceland, Ireland, Italy, Japan, Luxembourg, Mexico, the Netherlands, New Zealand, Norway, Poland, Portugal, South Korea, Spain, Sweden, Switzerland, Turkey, the United Kingdom, and the United States.

# Chapter 4
# Elements of a Research Study

## Planning

In the research process, particularly in regulated research, it is necessary to prospectively establish intended research through the development of a study plan or protocol. This is a very important document, typically authorized by a team of testing facility scientists (e.g., toxicologists, pathologists) who have the education and experience to appropriately conduct the type of research described.

The protocol, SOPs, and the GLPs provide the necessary documented requirements for the QAU monitors to audit the research study.

The GLPs are also a research-management tool, helping to monitor the planning, performance, and reporting stages of a study. In the regulatory environment, monitoring is a highly structured process of confirming the compliance of the planning, performance of procedures, and verification of the scientific report as reflected through study data documentation.

The regulations (21 CFR 58.120(a))* specify that each study must have an approved written protocol, or study

*The FDA GLPs are very typical and, therefore, have been routinely referenced and quoted from in this publication (FDA GLP 1979).

plan, that indicates the objectives and all methods for the conduct of the study. The study director, who has overall responsibility for the technical conduct of the study, must assure that the protocol is approved and, if appropriate, that it contains the following elements specified by the regulations (21 CFR 58.120(a)):

(1) A descriptive title and statement of the purpose of the study.
(2) Identification of the test and control articles by name, chemical abstract number, or code number.
(3) The name of the sponsor and the name and address of the testing facility at which the study is being conducted.
(4) The number, body weight range, sex, source of supply, species, strain, substrain, and age of the test system.
(5) The procedure for identification of the test system.
(6) A description of the experimental design, including the methods for the control of bias.
(7) A description and/or identification of the diet used in the study as well as solvents, emulsifiers, and/or materials used to solubilize or suspend the test or control articles before mixing with the carrier. The description shall include specifications for acceptable levels of contaminants that are reasonably expected to be present in the dietary materials and are known to be capable of interfering with the purpose or conduct of the study if present at levels greater than established by the specifications.
(8) Each dosage level, expressed in milligrams per kilogram of body weight or other appropriate units, of

the test or control article to be administered and the method and frequency of administration.

(9) The type and frequency of tests, analyses, and measurements to be made.

(10) The records to be maintained.

(11) The date of approval of the protocol by the sponsor and the dated signature of the study director.

(12) A statement of the proposed statistical methods to be used.

Accompanying each protocol, a schedule of activities or procedures is included that can be followed by the study director and technicians, and may be referred to by the QAU for scheduling audits. All changes or revisions to the protocol and the reasons for the changes must also be documented, signed by the study director, dated, and maintained with the protocol.

### Performance

Now that an approved protocol has been furnished and audited, the QAU manager will assign an auditor to monitor the performance of the study. In this active phase of the study, the auditor will determine whether the study is compliant with the following GLP requirements from the FDA regulations (21 CFR 58.130):

(a) The nonclinical laboratory study shall be conducted in accordance with the protocol.

(b) The test systems shall be monitored in conformity with the protocol.

(c) Specimens shall be identified by test system, study, nature, and date of collection. This information shall be located on the specimen container or shall accompany the specimen in a manner that precludes error in the recording and storage of data.

(d) Records of gross findings for a specimen from postmortem observations should be available to a pathologist when examining that specimen histopathologically.

(e) All data generated during the conduct of a nonclinical laboratory study, except those that are generated by automated data collection systems, shall be recorded directly, promptly, and legibly in ink. All data entries shall be dated on the date of entry and signed or initialed by the person entering the data. Any change in entries shall be made so as not to obscure the original entry, shall indicate the reason for such change, and shall be dated and signed or identified at the time of the change. In automated data collection systems, the individual responsible for direct data input shall be identified at the time of data input. Any change in automated data entries shall be made so as not to obscure the original entry, shall indicate the reason for change, shall be dated, and the responsible individual shall be identified.

In monitoring, the auditor may have observed, for example, that a scale or balance utilized for obtaining animal body weights had an expired calibration. An Inspection Report similar to the one shown in Figure 1, may be issued and processed for appropriate response and review by the study director and management.

## Figure 1 Inspection Report

### NORTHERN RESEARCH, INC.
### QUALITY ASSURANCE

#### INSPECTION REPORT

To: V. D. Holem, PhD and J. Beaver, DVM

STUDY DIRECTOR MONITOR

Please respond to the observation below, and return this report to the Quality Assurance Unit by 9/6/03.

Report Number: 03-092
Study Number: 02-147 Page: 1 of 1
Test Article: ST-205 Date: 8/6/03

This is a CONFIDENTIAL legal document that is to be maintained by Quality Assurance for the period of time as prescribed by the Good Laboratory Practice regulations.

STUDY PHASE(S) INSPECTED INSPECTION DATE(S)

| | | |
|---|---|---|
| Inspection: | | 5/24/03, 6/12/03, 6/20/03, 7/8/03, 8/2/03 |
| In-process: | √ | |
| Completion: | | |
| Reporting: | | |
| | D. C. Holst | D. C. Holst 8/6/03 |

OBSERVATION: The scale used in measuring body weights had an expired calibration date. Weighings were done 8/2/03; the labeled calibration expiration date of this scale, Asset #434, was 7/25/03.

RESPONSE: All the scales and balances within this facility are in a scheduled calibration program. The Calibration Record for the above scale indicates that it was calibrated on 7/24/03; however, the calibrator had failed to replace the calibration tag with one having the correct information. Metrology has been contacted and will make the correction.

| | | | |
|---|---|---|---|
| Prepared By: | M. H. Soukup | Date: | 8/8/03 |
| Approved By: | J. Beaver | Date: | 8/9/03 |
| Proposed Completion Date(s): | 8/14/03 | | |
| Study Director: | V. D. Holem | Date: | 8/12/03 |

QAU REVIEW:

| | |
|---|---|
| Action Plan Acceptance Date: | August 13, 2003 |
| By: | D. C. Holst |
| Completed: | 8/19/03 |

### Reporting

The FDA regulations (21 CFR 58.185(a)) require that a final scientific report be prepared for each nonclinical laboratory study and that it shall include the following:

(1) Name and address of the facility performing the study and the dates on which the study was initiated and completed.
(2) Objectives and procedures stated in the approved protocol, including any changes in the original protocol.
(3) Statistical methods employed for analyzing the data.
(4) The test and control articles identified by name, chemical abstracts number or code number, strength, purity, and composition or other appropriate characteristics.
(5) Stability of the test and control articles under the conditions of administration.
(6) A description of the methods used.
(7) A description of the test system used. Where applicable, the final report shall include the number of animals used, sex, body weight range, source of supply, species, strain and substrain, age, and procedure used for identification.
(8) A description of the dosage, dosage regimen, route of administration, and duration.
(9) A description of all circumstances that may have affected the quality or integrity of the data.
(10) The name of the study director, the names of other scientists or professionals, and the names of all supervisory personnel, involved in the study.

(11) A description of the transformations, calculations, or operations performed on the data, a summary and analysis of the data, and a statement of the conclusions drawn from the analysis.
(12) The signed and dated reports of each of the individual scientists or other professionals involved in the study.
(13) The locations where all specimens, raw data, and the final report are to be stored.
(14) The statement prepared and signed by the quality assurance unit as described in Sec. 58.35(b)(7).

It is also most important that the final report be signed and dated by the individual who has had the overall technical responsibility for the research, the study director.

# Chapter 5
# Applying Good Laboratory Practice

## The Quality Assurance Unit

In the regulations, confirmation of a research study's compliance has been delegated to the QAU. This is an independent function that monitors each study to assure management that the facilities, equipment, personnel, methods, practices, records, and controls conform with the regulations. It may reside in one of several areas of an organization (e.g., Quality Assurance, Regulatory Affairs, or Research & Development), but it must be kept independent of management directly associated with the nonclinical study. The QAU may consist of a manager and one or more auditors, such as displayed in Figure 2.

The auditors inspecting a regulated study should direct their attention toward observing the professionals' or technicians' compliance with the protocol and SOPs, and toward the proper documentation of the event and the ensuing measurements as raw data. From these data, a statistical report will be written or integrated into the scientific report. It will be the auditor's responsibility to audit the interim or final scientific report, confirming that the raw data is accurately portrayed.

## Standard Operating Procedures

The QAU should have written SOPs for purposes of consistency and accountability, and as a training aid. Aside from

**Figure 2 The Quality Assurance Unit (QAU)**

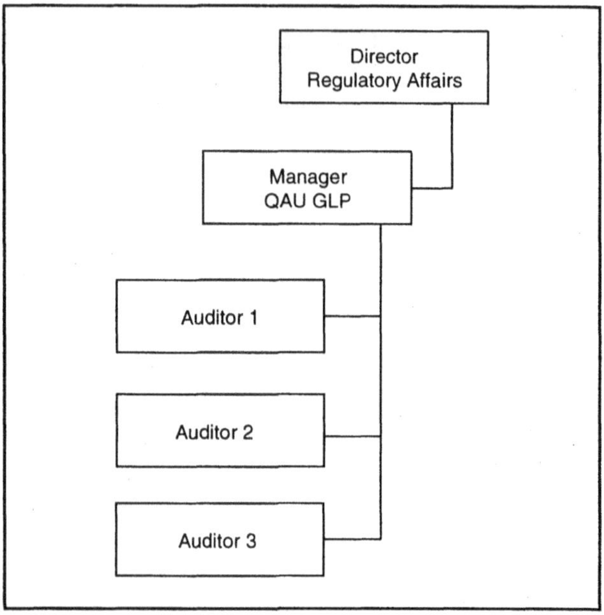


their regulatory necessity, SOPs are a must for management. The use of approved SOPs results in procedural consistency and serves as a valuable tool in replicating study parameters in any reevaluation of a given study. They serve as an excellent reference in orienting and training new personnel. The regulations specify that the QAU maintain SOPs that address their responsibilities, the records maintained, and the method of indexing such records. Some SOP titles may include the following:

- Records Maintained by the Quality Assurance Unit
- Maintenance of the Master Schedule
- Maintenance of Study Protocols
- Monitoring Nonclinical Studies

- System Auditing
- Scientific Report Auditing
- Preparation of the Quality Assurance Unit Statement
- Reports to Management and the Study Director
- Training Study-Associated Personnel in GLP Regulatory Requirements

The testing facility has record retention requirements described in the regulations, but the QAU is specifically responsible for the retention of some additional ones. These include the Master Schedule and copies of all protocols pertaining to the regulatory studies. The time of retention of these documents may vary depending on the regulatory agency's decision regarding the research and marketing application (e.g., retention requirements differ between the FDA, the EPA, the OECD, or a specific country's GLP standard); therefore, the specific regulations should be referenced.

### Inspections and Audits

One of the QAU's primary functions is monitoring each regulated study to assure management that the facilities, equipment, personnel, methods, practices, records, and controls conform with GLP. Study protocols, or plans, are usually issued by the research department responsible for the study, which furnishes a copy to the QAU prior to initiating active phases of the study. It is through this type of communication that the QAU can establish and maintain the required Master Schedule, as shown in Figure 3. Keyed to the schedule are specific projected events, protocol issuance, inspec-

## Figure 3    Master Schedule

Page 1
STATUS: March 27, 2004

NORTHERN RESEARCH, INC.
QUALITY ASSURANCE

**MASTER SCHEDULE**

**PROJECTIONS:**

**QAU Inspections** I

**Protocol Issued** P

**Final Report** R

**Interim Report** r

| | | | | | | 2004 | | | | | | | | | | 2005 | |
|---|---|---|---|---|---|---|---|---|---|---|---|---|---|---|---|---|---|
| **Test Article** | **Study Identity** | **Test System** | **Study Initiation** | **Study Director** | **Nature and Current Status Final Report** | **M** | **A** | **M** | **J** | **J** | **A** | **S** | **O** | **N** | **D** | **J** | **F** |
| AC-123 | 03-322 | Mouse | 1/07/04 | Smith | Acute PO Tox/In Preparation | R | | | | | | | | | | | |
| AC-123 | 04-107 | Rat | 3/09/04 | Foster | One-Month IV Tox/In-process | I | I | | R | | | | | | | | |
| BD-149 | 03-257 | Dog | 9/19/03 | Foster | Three-Month IV Tox/Issued 3/26/04 | R | | | | | | | | | | | |
| BD-149 | 03-314 | Monkey | 12/17/03 | Williams | Six-Month IV Tox/In-Process | I | | | I | | | | R | | | | |
| RS-160 | 04-059 | Rat | 3/25/04 | Smith | Acute IVTox/Protocol | P | I | R | | | | | | | | | |
| RS-160 | 04-103 | Rat | 4/09/04 | Smith | One-Month IV Protocol | | P | I | | R | | | | | | | |
| TM-172 | 03-337 | Salmonella | 3/05/04 | Gentry | Ames Test/In-Preparation | I | R | | | | | | | | | | |
| VW-178 | 03-157 | Dog | 8/20/03 | Foster | Twelve-Month Diet/In-Process | | | I | | | I | | | | | | R |
| ST-205 | 02-145 | Mouse | 08/06/02 | Williams | Two-Year Carcinogenicity Diet/ In-Process | r | | I | | | I | | | I | | | I |
| ST-205 | 02-147 | Rat | 09/20/02 | Holem | Two-Year Carcinogenicity Diet/ In-Process | I | r | | I | | | I | | | I | | |
| WS-310 | 03-258 | Dog | 10/02/03 | Anderson | Three-Month PO Tox/ In Preparation | | R | | | | | | | | | | |

tions, report completion, and audits. Some of the GLP Master Schedule requirements include the following:

- Indexing by test article
- Identities of the test system
- Nature of study
- Date study was initiated
- Current status of each study
- Identity of the sponsor
- Name of the study director

The QAU also maintains documentation of each inspection in the Study Inspection Record. This record includes the date of inspection, the study inspected, the phase of the study inspected, the auditor, findings and problems, and corrective actions recommended, as illustrated in Figure 4.

Significant problems that could affect the study's integrity should be reported to the study director or to management immediately. Management may be informed of these observations by submitting a copy of the Inspection Reports to their attention. After corrective actions have been recommended and completed, the Inspection Reports are circulated to management for their information and comment.

After the active phases of the study have been completed, the study director coordinates the development of a written scientific report, which should reflect the raw data and scientific findings of the study. The scientists and technicians have persevered for months or years, carefully contributing their expertise to a study. Although they may have diligently tried to maintain protocol compliance, there may be events that were not documented or records that differ. The QAU

## Figure 4 Study Inspection Record

Version 2

**NORTHERN RESEARCH, INC.**
**QUALITY ASSURANCE**

**STUDY INSPECTION RECORD**

Study Title: Three-Month PO Toxicity
Test Article: WS-310
Study Identity: 03-258
Study Species: Dog
Route: PO
Study Director: B. E. Anderson, PhD

| STUDY PHASE | PROCEDURE | INSPECTION DATES | SIGNATURE/ DATE | INSPECTION REPORT NO./DATE | CORRECTIVE ACTION ACCEPTANCE DATE |
|---|---|---|---|---|---|
| INITIATION | PRO | 10/2/03 | P. Verke 10/2/03 | | |
| | AWS | 10/2/03 | P. Verke 10/2/03 | | |
| | FDR | 10/9/03 | P. Verke 10/9/03 | | |
| IN-PROCESS | AWS | 10/7/03 | P. Verke 10/7/03 | | |
| | DOS | 10/9/03 | P. Verke 10/9/03 | 03-097 10/9/03 | 10/17/03 |
| | FDR | 11/12/03 | P. Verke 11/2/03 | | |
| | BCN | 12/2/03 | P. Verke 12/2/03 | | |
| | OBS | 12/6/03 | P. Verke 12/6/03 | | |
| | DOS | 12/10/03 | P. Verke 12/10/03 | | |
| COMPLETION | NEC | 1/2-3/04 | P. Verke 1/3/04 | | |
| | CCH/HEM | 1/2-7/04 | P. Verke 1/7/04 | 04-002 1/7/04 | 1/23/04 |
| | MOC | 2/19/04 | P. Verke 2/17/04 | | |
| REPORTING | RPT | 4/7-9/04 | P. Verke 4/9/04 | | |
| | | 4/16/04 | P. Verke 4/16/04 | 04-017 4/16/04 | 4/27/04 |

auditor's responsibility is to carefully review these study source documents and to assure that the report accurately reflects these raw data and describes the methods and SOPs that were applied at the time of the study.

At the completion of the audit, the QAU manager will issue what is commonly referred to as the QAU Statement. This record, illustrated in Figure 5, includes the dates inspections were made and the dates findings were reported to management and the study director. It is typically appended to the scientific report so that it is readily available for reference by the regulatory agency.

The QAU also maintains the documentation supporting an inspection or audit. This is a very structured process, one in which properly signed records are developed that include the date of inspection, the study inspected, the phase or segment of the study inspected, the auditor, findings and problems, corrective actions recommended, and, if appropriate, the scheduled date of reinspection. Examples of documents providing this type of information are shown in the Study Inspection Record and the Inspection Report. Two separate reports are preferred because the Study Inspection Record may be requested by and reviewed by the agency during their inspection. As you will note, this record includes the study phases, inspection dates, auditors' signatures, and dates. The GLPs specify that this shall be made available to authorized individuals of the agency. The Inspection Report, however, containing findings, problems, and corrective actions, is a confidential report exempt from routine agency inspections.

All Inspection Reports are submitted to the study director for a coordinated response. Depending on the observation, it may be necessary for several respondents to provide an inte-

## Figure 5 QAU Statement

### NORTHERN RESEARCH, INC.

#### QUALITY ASSURANCE UNIT STATEMENT

Periodic Inspections of Study Number 03-257, entitled "Three-Month IV Toxicity Study of BD-149 in Dogs," were conducted on all phases by the Quality Assurance Unit of Northern Research, Inc.

The study was inspected on the following dates:

September 19, 20, 24, 25, 2003
October 8, 15-17, 23-25, 2003
November 12-15, 2003
December 17-20, 2003
January 13, 14, 2004
February 5, 28, 2004
March 4, 6, 2004

Results of inspections were reported to management and the study director on 10/25/03, 12/20/03, and 3/6/04.

To the best of the signator's knowledge, there were no significant deviations that affected the quality or integrity of the study from the Good Laboratory Practice regulations.

B. Victoria Almquist ______ 3/6/04 ______

B. Victoria Almquist — Date
Manager, Quality Assurance

grated corrective action. The QAU auditor, for example, may have observed that the statistics of a clinical chemistry table or summary were in error. The study director must be a respondent, as shown in the Inspection Report in Figure 6; additionally, it may be appropriate to require responses from the clinical pathologist and the study statistician.

## System Audits

Much of the inspection and auditing activities required by the GLPs are routinely oriented around the study protocol. Because of the nature of the specific studies being conducted, some aspects of the GLPs may seldom be challenged or evaluated, thus providing the need for a systems or maintenance audit. System audits have been encouraged as supplements to study-specific audits, providing a more in-depth inspection of the study processes in place of quick, spot checks of individual operations of a study.

By design, the scope of these audits provides a more thorough examination of personnel qualifications, training, facilities, equipment, data collection systems (both manual and computer oriented), and SOPs. Supportive departments may be audited thoroughly, using their SOPs as one of the guides in evaluating compliance.

Here are some of the compliance issues that may surface more readily through system audits of departments or functional groups:

- The reagents within a clinical chemistry laboratory may be used infrequently for some analyses. Have those

## Figure 6 Inspection Report

**NORTHERN RESEARCH, INC.**
**QUALITY ASSURANCE**

**INSPECTION REPORT**

To: J. B. Foster, PhD and L. A. McCarley, MS

STUDY DIRECTOR MONITOR

Please respond to the observation below, and return this report to the Quality Assurance Unit by 4/6/04:

Report Number: 04-027
Study Number: 03-257
Test Article: BD-149

Page: 1 of 1
Date: 3/6/04

This is a CONFIDENTIAL legal document that is to be maintained by Quality Assurance for the period of time prescribed by the Good Laboratory Practice regulations.

STUDY PHASE(S) INSPECTED — INSPECTION DATE(S)

| | | |
|---|---|---|
| Inspection: | | 2/28/04, 3/4/04, 3/6/04 |
| In-process: | | |
| Completion: | | |
| Reporting: | √ | |
| | A. D. Moulchin | *A. D. Moulchin 3/9/04* |

OBSERVATION: The following were observed in the appended statistical report:

In Table 2, Urinalysis, the mean pH of the $T_2$ females and $T_3$ males are respectively recorded as 6.5 and 7.3. Calculations associated with the raw data indicate respective mean pH of 6.7 and 7.0

In Table 5, Absolute and Relative Animal Organ Weights, the mean kidney weight of the $T_4$ males is recorded as 44.23 g. Calculations associated with the raw data indicate a mean of 47.57 g.

RESPONSE: The errors noted above in Tables 2 and 5 of the Statistical Report have been corrected and the database changed to reflect the raw data.

| | | | |
|---|---|---|---|
| Prepared By: | *F. B. Christ, PhD (Statistician)* | Date: | *3/12/04* |
| Approved By: | *R. B. Schubert, PhD (Stats. Mgr.)* | Date: | *3/17/04* |
| Proposed Completion Date(s): | *3/20/04* | | |
| Study Director: | *J. B. Foster* | Date: | *3/18/04* |

QAU REVIEW:

| | |
|---|---|
| Action Plan Acceptance Date: | *3/20/04* |
| By: | *A. D. Moulchin* |
| Completed: | *3/23/04* |

with expired dating been used? Are there any unused with expired dating?

- A newly installed unit for diet mixing is being contemplated for use in several studies. Have the mixing procedures and the equipment cleaning procedures been documented and validated?
- In the various laboratories, have the equipment or instrument maintenance logs been utilized, and are they on schedule, identifying dates of repair and cleaning?
- Have the histology slide or wet tissue storage areas been evaluated recently? Have slides or tissues been checked out recently? Are departments and groups following SOPs?
- Have all the nonclinical statistical computer programs been validated? Is there supporting documentation?

In conducting this audit, initially request copies of the department's SOP index. From this list, select SOPs that are considered critical for nonclinical studies. If this is a supportive department, such as biometrics, the computer programs associated with generating randomization schedules or statistical results would be good choices. Select several studies of different types: acute, subchronic, chronic, and TFR (teratogenic, fertility, and reproductive) studies. By referring to the appropriate SOP and the specific scientific reports, you may wish to proceed as follows:

- Determine if the randomization and statistical software programs used were properly validated.
- Does timely documentation exist supporting the verifi-

cation of the computer program versions referred to in the scientific reports?

- Do test sets of data exist?
- Have the results been retained and do these match the expected values?
- If third-party computer programs are used, are they appropriately licensed and are they being supported?
- Are changes in software controlled, documented, and approved appropriately according to the SOPs?

SOPs in other departments can also be examined for compliance. Reagents within a clinical chemistry laboratory may be used infrequently for some analyses. Some reagents, although not used, may have expired, and should be discarded. Is the department following the SOP maintaining the availability of currently accepted reagents and appropriately discarding those that have expired?

After completing an audit, it is always helpful for the chairperson to schedule a meeting with the audit team and review their observations. There may be observations that the auditing team may have that would more appropriately be mentioned in the closeout meeting, but would not necessarily be documented. Identify for the group those observations that should be drafted into the Audit Report, as illustrated in Figure 7.

The Audit Report should include an audit summary, and, of course, the audit observations, client responses, targeted corrective-action date commitments, corrective-action completion date, and management's plan with accompanying approval. The QAU should review the responses and decide upon their acceptance or rejection. Those that are recog-

## Figure 7 Audit Report

**NORTHERN RESEARCH, INC.**
**QUALITY ASSURANCE**

**AUDIT REPORT**

| | | |
|---|---|---|
| Report Number.: | 03-105 | Page 1 |
| Date: | 11/26/03 | |
| Department Audited: | Analytical Chemistry | |
| Location: | Coeur d'Alene, ID | |
| Date(s) Audited: | 11/18-21/03 | |
| Auditors: | T. Nelson | T. Nelson, 11/26/03 |
| OBSERVATION: | 1. The Test Method validation file for determining the potency of test article ST-211 had been approved by the department manager on 10/16/03. Animals in study 03-125 were dosed with this compound seven days prior to approval of the validated test method on 10/9/03.<br><br>2. SOP 796-35.02, MS-345 Determination-HPLC, approved 5/3/01, requires that for each standardization and test, the chromotographic system components be identified on the charts.<br><br>Examination of the applicable HPLC charts, run 11/06/03, indicate that only the column has been identified. | |
| RESPONSE: | 1. The Test Method for the determination of potency for ST-211, includes more than three determinations consistently using the same methodology and performed by the same scientist. This had been reviewed in Dr. Shaw's laboratory notebook, accepted and acknowledged by management on 10/3/03. The formal, Test Methodology Report was signed, as indicated, 10/16/03. An explanation will be written, filed both in the validation file and with the documentation for study 03-125.<br><br>2. These identity requirements had inadvertently been excluded. However, they had been recorded in the scientist's laboratory notebook, and have now been transcribed to the proper charts. This individual and other members of the department have been cautioned to be more meticulous. | |

| | | | |
|---|---|---|---|
| Prepared By | M. G. Aspatore, PhD, Methods Development | Date: | 11/28/03 |
| Approved By: | R. A. Bloch, PhD, Laboratory Manager | Date | 12/4/03 |

QAU REVIEW:

| | | | |
|---|---|---|---|
| Report Return Date: | 12/5/03 | | |
| Accepted By: | B. V. Almquist | Plan Accepted Date: | 12/6/03 |
| Title: | Manager QA | | |
| Auditor: | T. Nelson | Plan Accepted Date: | 12/18/03 |

nized as significant observations with accompanying corrective actions should be scheduled for an audit follow-up and subsequently documented, if appropriate, as completed.

### Contract Laboratory Audits

At some companies, it is policy to contract out all nonclinical safety studies; at others, contracting out studies may be done only when the need arises. The reasons for contracting varies. Animal facilities may be near or at capacity; a temporary shortage of professional help may be contemplated; the study may be unique, requiring special techniques, equipment, or environments; or the biohazards of the compound or positive control may make it desirable to test it away from the sponsor's facilities.

For studies that are planned with contract laboratories, the responsible sponsor or scientific monitor should inform the QAU of his or her intended study plan and request a GLP evaluation of the contractor. Obtain a copy of the protocol and any brochures or literature describing the contract laboratory–its operations and personnel responsibilities.

Contact the contract laboratory's QAU manager, and advise him or her of the contemplated GLP study at his or her company. Schedule a visit, clarifying that it will be an evaluation audit of their facilities, operations, SOPs, and QAU. At the contractor's site a brief kickoff meeting is usually held with various members of the contractor's management. This provides them with a better understanding of the scope of the audit, and allows them to recommend changes in your agenda to facilitate a more efficient or comprehensive audit.

After this preliminary meeting, you should meet with the contractor's QAU manager and the tentative study director. Advise them of the type of documentation to be reviewed. This might include: the curriculum vitae of any professional personnel engaged in the study, along with those of supervision and study technicians; the SOPs associated with the facility's operations; and those SOPs addressing the specific type of study being contemplated and required by the GLPs. A thorough review should also be done of the QAU's SOPs and their personnel's curriculum vitae. This is extremely important, because in your absence, their QAU will be delegated the responsibilities for study inspection and audits.

The QAU manager should provide a tour of the facilities, especially those areas associated with your study interests. Bring up concerns immediately, and note them accordingly. Some queries that may be addressed are:

- Are these studies being conducted in separate rooms?
- If biohazardous materials are being tested, are the air handling systems separate, properly filtered, using 100 percent makeup air?
- How will they handle any protocol requirements regarding periodic analyses of test mixtures?
- Is the test methodology validated, the documentation available and compliant?
- How are the various containers and mixing devices cleaned?
- Is the adequacy of cleaning determined and validated?
- Is the autopsy room designed for multiple necropsies, with adequate space, providing separation, and is it adequately equipped?

### Training

QAU personnel engaged in inspection, auditing, and administrative GLP responsibilities come from a variety of career experiences and educational backgrounds. Perhaps all have had scientific training–baccalaureate, masters, or doctoral degrees. When the GLPs were first proposed, one of the industry's immediate concerns was that there would be a shortage of personnel qualified to fulfill these responsibilities. Management felt that to meet their perceived qualifications, it would be necessary to assign scientists already productive and familiar with their particular area of science. The final regulations, however, elaborated further and clearly defined the function of the QAU as one assuring compliance with procedural and administrative requirements rather than overseeing the technical aspects of study conduct. QAU personnel need not be limited to professionals or scientists, but these certainly remain major assets in this field and are recommended prerequisites. Also, some sponsors have encouraged their auditors to attain certification through an existing program established by the American Society for Quality (ASQ). This society, through confirmation of acceptable experience and testing, may certify an individual as a Certified Quality Engineer or Certified Quality Auditor. Similarly, one of this industry's major associations, the Society of Quality Assurance (SQA), has established a professional certification for members professionally engaged in GLP regulatory responsibilities. Regardless of these credentials, auditors must continue to demonstrate, through interaction with their peers at QAU training programs, seminars, and conferences, a continued understanding and sustained record of appropriate interpretations of these regulations.

A training aid that has been very beneficial and that supports the auditing process, is the checklist. The checklist provides a basis of improved audit consistency; it serves as a supportive auditing document; and it is particularly helpful for new auditors. For example, with regard to the GLP requirements of the study protocol, the auditor may have reference to the list of requirements described in the checklist partially illustrated in Figure 8. Many of the requirements are common to other standards or principles, and these, too, can be incorporated in a single list. In this instance both the FDA GLPs, as well as those of the OECD (OECD 1997) are referenced. Checklists can be developed for the entire spectrum of GLPs, serving as excellent auditing documentation and training aids (Anderson 2000).

The GLPs define the QAU's regulatory range of duties, but some management may prefer increasing the scope of the assignment, including auditing of the technical aspects of the study. Because of associated company functions that may be assigned (for example, archiving, calibration, regulatory affairs, and GCPs), personnel qualifications may vary considerably between different organizations. Within a QAU, for example, there may be three or four levels of job responsibilities, some including only inspection, others inspection, auditing, or administration. After being assigned to the QAU, qualified individuals should proceed with pursuing their training needs. Depth of training may vary depending on job level, responsibilities, and prospective career paths. Attending external seminars on GLP topics, as well as other subjects enhancing the auditor's understanding of the client

## Figure 8 Portion of a GLP Checklist

| GLP AUDIT CHECKLIST | Response: Yes | No | N/A | References: 21 CFR | OECD |
|---|---|---|---|---|---|
| **T. Protocol for and Conduct of a Nonclinical Laboratory Study** | | | | **58.120** | |
| 1. Does each study have an approved written protocol that clearly indicates the objectives and all methods for the conduct of the study? | ☐ | ☐ | ☐ | | 8,8,1(1) |
| 2. Does the protocol contain, as applicable, the following information: | | | | | |
| 2.1 A descriptive title and statement of the purpose? | ☐ | ☐ | ☐ | | 8,8.2(1)(a)&(b) |
| 2.2 Identification of the test and control article (TCA) by name, chemical abstracts number, or code number? | ☐ | ☐ | ☐ | | 8,8.2(1)(c) |
| 2.3 The name of the sponsor and the name and address of the testing facility at which the study is being conducted? | ☐ | ☐ | ☐ | | 8,8.2(2)(a)&(b) |
| 2.4 The number, body weight range, sex, source of supply, species, strain, substrain, and age of test system? | ☐ | ☐ | ☐ | | 8,8.2(5)(b) |
| 2.5 The procedure for identification of the test system? | ☐ | ☐ | ☐ | | 8,8.2(5)(b) |
| 2.6 A description of the experimental design, including the methods for control of bias? | ☐ | ☐ | ☐ | | 8,8.2(5)(e) |
| 2.7 A description and/or identification of diet used in the study as well as solvent, emulsifiers, and/or other materials used to solubilize or suspend the TCA before mixing with the carrier? | ☐ | ☐ | ☐ | | |

group's studies, would be beneficial. For GLP inspection or auditing, in-depth scientific knowledge may not be necessary, but familiarity is certainly essential. Internal seminars can also be organized by requesting brief presentations from your own organization's professionals in their field of expertise. A series of one- or two-hour sessions is common, and if these seminars are recorded by video or audio, new employees may readily access the information for an efficient, continuing training program. The intended syllabus should be scheduled in a goal-oriented program for each individual. Documentation supporting satisfactory completion of this training should include confirmation by management, and, if appropriate, acceptance of the individual as a qualified GLP auditor.

# CHAPTER 6
# Internet Resources

The Internet has become one of the most useful tools in research. Many of the governmental agencies throughout the world are now providing convenient Internet access to their regulations, guidelines, and other supplementary information associated with the GLPs. This immediate access to timely information should provide research, quality assurance, and regulatory functions a platform for a more expeditious and accurate processing of studies.

Listed here in Subsection A, Organizations, are Web sites that should offer a variety of helpful references. Besides the regulatory agencies' Uniform Resource Locator (URL) addresses, other relevant organizations, societies, and academic institution addresses are included. Although the Internet is a relatively new medium, it has generated a vast increase in associated vernacular terms, some of which are explained in Subsection B, Internet Terms.

Many of the sites contain hypertext formatting, allowing the user to visit an enormous number of sites in a relatively short time. The government agency sites are organized particularly well, along with those of the SQA and the University of Georgia (UGA). The hypertext entries should be extremely beneficial and will allow the researcher to expeditiously acquire answers to a boundless variety of research and regulatory questions.

## A. Organizations

***American Association for Laboratory Animal Science (AALAS)***

http://www.aalas.org

This organization of more than ten thousand clinical veterinarians, technicians, technologists, educators, business people, researchers, administrators, animal producers, national and international experts is the premier forum for the exchange of information and expertise in the care and use of laboratory animals. They are dedicated to the humane care and treatment of laboratory animals and the quality research that leads to scientific gains that benefit mankind and animals.

***American Association for the Advancement of Science***

http://www.aaas.org

Founded in 1848, this association has a large enrollment, now serving ten million members. Their mission includes the advancement of science for the benefit of all people through: furthering the work of scientists and facilitating scientific cooperation; fostering scientific freedom and responsibility; promoting human welfare through science; advancing education in science; and increasing public understanding and appreciation of the importance and promise of science for human progress.

***Association for Assessment and Accreditation of Laboratory Animal Care (AAALAC International)***

http://www.aaalac.org

This organization promotes the humane treatment of animals in science through a voluntary accreditation program. Many companies, universities, hospitals, government agencies, and other research institutions have earned AAALAC accreditation, demonstrating their commitment to responsible animal care and use. These institutions participate in AAALAC's program, and comply with local, state, and federal laws that regulate animal research.

***CDC (Centers for Disease Control and Prevention)***

http://www.cdc.gov

This homepage provides a general introduction to the CDC. It includes several hyperlinks to other CDC pages, including: About CDC; In the News; Travelers' Health; Health Information; Publications; Software & Products; Data & Statistics; Training & Employment; and Funding.

***EPA (Environmental Protection Agency)***

http://www.epa.gov

This homepage serves as a launching site for various informational pages, including: About EPA; News & Events; Offices, Labs & Regions; Projects & Programs; Laws & Registrations; Publications; Databases & Software; Money Matters.

http://www.epa.gov/irmpoli8/irm_galp

An electronic version of the document entitled "EPA Directive 2185: Good Automated Laboratory Practices (GALP)." It includes hypertext links to the following related pages: Introduction; GALP Overview; and GALP Implementation Assistance.

http://epa.gov/epahome/rules.html

Included in these documents are: Regulations & Proposed Rules; Codified Regulations; Current Legislation; and

Laws. These subjects, along with many other helpful entries, are provided in hypertext for immediate access to other Web pages.

http://www.ovpr.uga.edu/qau/advisor2.html

This electronic "Index to the EPA GLP Advisories" is numbered and the various index entries may be readily accessed through hypertext linkage.

***FDA (Food and Drug Administration)***

http://www.fda.gov

This is the FDA's homepage, providing an extensive range of subjects, all with hypertext entries to other pages in the FDA's Web site. Included are: Biologics; Animal Drugs; Cosmetics; Medical Devices/Radiological Health; Freedom of Information; Field Operations; Regulations & Information; Toxicology; and Medical Products Reporting.

http://www.fda.gov/cder/index.html

This is the homepage for the Center for Drug Evaluation and Research. It provides hypertext links to further information about CDER, Drug Information, and Regulatory Guidance.

http://www.eduneering.com/fda/courses/fdatour/welcome.html

This is the FDA's welcome page, giving you an option to take *"A Tour of FDA."* It also introduces you to the availability of many Web-based courses offered as part of the FDA's Office of Regulatory Affairs University (ORA U).

http://www.fda.gov/cvm

This is an introductory page to the Center for Veterinary Medicine's (CVM) portion of the FDA Web site. It describes the scope of the CVM's responsibilities and includes links to other CVM Web pages, including: Important News; On-Line Library; The Green Book; Links; and About CVM.

http://www.ovpr.uga.edu/qau/prefda1.html

This Web page contains an electronic version of the GLP Preamble. A hypertext index is provided for GLP (21CFR 58), covering the full range of subjects with numbered comments.

http://www.ovpr.uga.edu/qau/fdaglp_a.html

This is an index to the GLP (21 CFR 58) provided by UGA's QAU. Each Section (e.g., 58.31 Testing facility management) is formatted in hypertext, allowing convenient access to the site containing the specific Section's text.

http:///www.cgmp.com/htmfiles/211.htm

This is the hypertext index for the current Good Manufacturing Practice (cGMP) regulations, finished pharmaceuticals (21 CFR 211).

http:///www.fda.gov/cder/dmpq/cgmpnotes.htm

This is an internal FDA document providing access to cGMP issues in a timely manner. It also includes updates on cGMP projects and provides help in clarifying existing policy.

http://www.fda.gov/opacom/hpnews.html

FDA News and Publications is a source of information about Recent Press Releases and Talk Papers. It also has topics on Current Issues of Other FDA Publications. These are all linked with the Internet. References include an Index and Archives.

http://www.access.gpo.gov/cgi-bin/cfrassemble.cgi?title=21

Title 21, Code of Federal Regulations. This provides access to all volumes and Parts of Title 21.

***Institutional Animal Care and Use Committee (IACUC)***

http://www.ovpr.uga.edu

http://www.wsu.edu

These are the respective sites of the University of Georgia and Washington State University, providing information associated with their IACUCs, in conjunction with the US Department of Agriculture Animal Welfare Act. Also, within this section, refer to the appropriate USDA hyperlink.

***Interpharm Press***

http://www.interpharm.com

Interpharm Press specializes in publishing books in the area of applied technology and regulatory affairs, with primary audiences of professionals in the pharmaceutical, medical device, diagnostic, and biotechnology manufacturing industries. It is the world's premier provider of technical books, pharmacopoeias, software, regulatory documents, and audiovisual training tools for professionals.

***The Library of Congress***

http://lcweb.loc.gov/homepage/lchp.html

This is the library's homepage, allowing convenient hypertext access to related Web pages on many topics. Perhaps of primary interest would be the Library Services, which includes resources for libraries, information professionals, and researchers. These include: Acquisitions, Cataloging, Preservation, Research, Special Programs, Standards, and access to the catalogs of the Library of Congress and other libraries.

***National Archives and Records Administration***

http://www.access.gpo.gov/nara/cfr/index.html

This site provides access to the latest published Code of Federal Regulations. Additionally, it has other important references formatted in hypertext, allowing convenient access to the specific Internet sites. These other references include: *Federal Register;* Privacy Act Issuances; Public Laws; United States Government Manual; Weekly Compilation of Presidential Documents; US Congress Information; and GPO (Government Printing Office) Access Search Page.

***National Institutes of Health (NIH)***

http://www.nih.gov

This is the homepage of the National Institutes of Health. It provides access to related pages on the NIH Web site, including: Health Information; Grants & Funding Opportunities; News & Events; Scientific Resources; Institutes, Centers & Offices; and an overview of NIH.

***Organisation for Economic Co-operation and Development***

http://www.oecd.org

This is the OECD's homepage, providing a platform for accessing titles, which include Activities, About OECD, Statistics, News & Events, Recruitment, Bookshop, Free Documents, and Search OECD Centres.

***Society of Toxicology***

http://www.toxicology.org

This is a professional organization of scientists from academic institutions, government, and industry, representing scientists who practice toxicology in the US and abroad. The society encourages the acquisition and utilization of knowledge in toxicology and aids in the protection of public health. It has a strong commitment to education in toxicology and to the recruitment of students and new members into the profession.

***The Society of Quality Assurance (SQA)***

http://www.sqa.org

This is the homepage of SQA, dedicated "to promote the quality assurance profession through the discussion and exchange of ideas." The professionals in the society are focused primarily on quality assurance–related requirements associated with the GLPs and GCPs worldwide. More than 1,200 SQA members participate in national conferences or regional chapter meetings. Besides those registered members from the United States, internationally listed are members from Australia, Canada, Denmark, England, France, Germany, Italy, Japan, Luxembourg, Mexico, Sweden, Switzerland, and the United Kingdom. This site provides hyperlinks to numerous regulatory-related documents of interest to all those involved in the GLPs.

***United States Department of Agriculture (USDA)***

http://www.usda.gov

This is the USDA's homepage, with links to sites including About USDA, News & Information, Opportunities, and Agencies.

http://www.nalusda.gov

This is the National Agriculture Library site.

http://www.ars.usda.gov

This is the Agriculture Research Service site.

http://www.aphis.usda.gov/ac/polmanpdf.html

This site provides an index to the Animal Welfare Act, Animal Care Policy Manual.

***US House of Representatives***

http://www.house.gov

This site has numerous hypertext entries that could be very helpful for your interests, including Member Offices, Committee Offices, Leadership Offices, Other House Organizations, Commissions and Task Forces, Annual Congressional Schedule, The Legislative Process, Roll Call Votes, and House Committee Hearing Schedules and Oversight Plans.

http://www.house.gov/science

The United States House of Representatives Committee on Science provides a good example of the government's involvement in various aspects of scientific development. Hypertext topics currently include Technology Transfer Challenges and Partnerships; Report on Genetically-Modified Plants; Report on the Assessment of the Benefits, Safety, and Oversight of Plant Genomics and Agricultural Biotechnology; Science Policy Study; and Science Education.

***US Senate***

http://www.senate.gov

This United States Senate site includes a platform for linking to a variety of interests: Learning About the Senate, Legislative Activities, Committees, Senators, Committee Hearing Schedule, Bill Search, and Today' Schedule.

## B. Internet Terms

### *ASCII*

American Standard Code for Information Exchange. It generally denotes the transmission of text without formatting, and it is independent of any specific type of font. It is a universal data code applied in communications associated with minicomputers and personal computers. It consists of 128 possible character combinations (e.g., including upper- and lower-case letters of the alphabet, numerals 0 through 9, punctuation marks, and so on).

### *Cyberspace*

The universe of information available from the computer networks.

### *Digital Subscriber Line (DSL)*

A method for moving data over regular phone lines. DSL circuitry is much faster than a regular phone connection. The DSL circuits must be configured to connect two specific locations, similar to a leased line; however, the DSL configuration is not a leased line.

### *Domain*

Represented in the address (URL) by the designated group you wish to communicate with:

.com (Commercial site)
.edu (Educational institution site)
.gov (Government site)

.int (International site)
.mil (Military site)
.net (Network site)
.org (Organization site)

***Domain Name***

It is the name assigned to your ISP (Internet Service Provider, e.g., aol.com, iquest.net, Microsoft.com).

***e-mail***

Electronic mail. Messages sent from one person or organization to another via computer networks.

***Homepage***

This is the introductory page of a World Wide Web site, generally formatted with hyperlinks connecting the user to other related pages and sites. It should be noted that in my review of this Internet topic, this word is spelled inconsistently as a single word, homepage, or as two words, home page.

***Hyperlink***

This is a programmed connection between Web pages, usually highlighted or underlined. By clicking on it (may require simultaneous activation of the CONTROL key), the Web browser is activated, seeking out the requested page from another location of the Internet.

***Hypertext Markup Language (HTML)***

The standard coded programming language used for creating formatted web pages.

### *Internet*

The international network of networks that consists of two or more computers connected by a physical device, such as cable, or a wireless device, such as infrared or satellite wavelengths. The Internet is the biggest network the world has ever witnessed.

### *Internet Services Providers (ISP)*

These are organizations that provide access to the Internet by dialing in for this service. In order to access the Internet, it is necessary to connect through an ISP. Before choosing an ISP, it would be appropriate to have an experienced Internet user provide a recommendation. There are more than ten thousand ISPs at this time. You may find a listing and recommendation of some ISPs at Web address: http://thelist.iworld.com/.

### *Modem*

This is the acronym of "modulator-demodulator." It is an electronic device that permits computers to communicate with each other via telephone line, cellular signals, or television cables. The modem converts digital signals from a computer into analog signals that can be sent over the types of media referenced above. The receiving modem converts the analog signals back into digital ones that can be understood by the computer.

### *Search Engines*

These are useful tools utilized for seeking or searching for information on the Web. They each have different parameters in their search criteria and, therefore, provide a variety of source references. Some examples of search engines are Alta Vista (http://www.altavista.com), Excite (http://www.excite.com), Infoseek (http://infoseek.go.com), Lycos (http://www.lycos.com), Netscape Netcenter (http://www.netscape.com), and Yahoo (http://www.yahoo.com).

***Uniform Resource Locator (URL)***

This is the addressing system used in locating Web sites on the Internet. It is the protocol for identifying documents on the Web.

***Web Browser***

A software program that provides users with an interface in accessing and navigating the World Wide Web.

***Web Site***

Usually a collection of World Wide Web pages, consisting of a homepage and other linked pages.

***World Wide Web***

A computerized system of finding and accessing Internet files and programs utilizing hypertext. The World Wide Web allows users to navigate the Internet by following links from documents on one computer to documents on others. The Web employs html (the coding language of the World Wide Web), allowing files to be viewed in a graphical format.

# Conclusion

This, of course, was a very limited journey through the evolution and implementation of GLPs. Hopefully, this publication, as well as the others referenced, will motivate you further in your continued understanding and knowledge of GLPs.

Although these regulations were originally promulgated as a means of confirming the integrity and quality of research, the benefits have been more extensive, having ramifications worldwide. The good laboratory practice concept and movement has had major economic impact, potentially reducing research costs significantly. Studies that are confirmed and documented as conforming to the GLPs provides the foundation for eliminating the need for repeating or duplicating research studies. Policy such as this potentially allows regulatory agencies worldwide to consider in their review process GLP-conforming scientific reports as supportive of the sponsor's product application for approval. Perhaps most importantly, the savings in research time may provide a new therapeutic cure, benefiting the ultimate users, you or your loved ones.

# Glossary

The definitions that follow are generally those obtained from the FDA's 21 CFR Part 58, Good Laboratory Practice for Nonclinical Laboratory Studies. The listing below is not complete. Other agencies (e.g., Environmental Protection Agency 1983, 1984), or international organizations (e.g., Organisation for Economic Co-operation and Development 1997), may have additional, or similar, but different definitions. Some of these are included for your interest. It is suggested that the respective publications be referenced if a more extensive glossary of definitions is required, or that the publication *GLP Quality Audit Manual* (Anderson 2000) be referenced.

*Control Article* (FDA)

Any food additive, color additive, drug, biological product, electronic product, medical device for human use, or any article other than a test article, feed, or water that is administered to the test system in the course of a nonclinical laboratory study for the purpose of establishing a basis for comparison with the test article.

*Control Substance* (EPA)

Any chemical substance or mixture, or any other material other than a test substance, feed, or water that is administered to the test system in the course of a study for the purpose of establishing a basis for comparison with the test substance for known chemical or biological measurements.

*Nonclinical Laboratory Study* (FDA)

Any in vivo or in vitro experiments in which test articles are studied prospectively in test systems under laboratory conditions to determine their safety. The term does not include studies utilizing human subjects or clinical studies or field trials in animals. The term does not include basic exploratory studies carried out to determine whether a test article has any potential utility or to determine physical or chemical characteristics of a test article.

*Quality Assurance Unit* (FDA)

Any person or organizational element, except the study director, designated by testing facility management to perform the duties relating to quality assurance of nonclinical laboratory studies.

*Raw Data* (FDA)

Any laboratory worksheets, records, memoranda, notes, or exact copies thereof, that are the result of original observations and activities of a nonclinical laboratory study and are necessary for the reconstruction and evaluation of the report of that study. In the event that exact transcripts of raw data have been prepared (e.g., tapes that have been transcribed verbatim, dated, and verified accurate by signature), the exact copy or exact transcript may be substituted for the original source as raw data. Raw data may include photographs, microfilm or microfiche copies, computer printouts, magnetic media, including dictated observations, and recorded data from automated instruments.

*Raw Data* (OECD)

All original laboratory records and documentation, or verified copies thereof, that are the result of the original observations and activities in a study.

*Reference Substance* (OECD)

Any well-defined chemical substance or any mixture other than the test substance used to provide a basis for comparison with the test substance.

*Specimen* (FDA)

Any material derived from a test system for examination or analysis.

*Specimen* (OECD)

Any material derived from a test for examination, analysis, or storage.

*Standard Operating Procedures* (OECD)

Written procedures that describe how to perform certain routine laboratory tests or activities normally not specified in detail in study plans or test guidelines.

*Study* (OECD)

An experiment or set of experiments in which a test substance is examined to obtain data on its properties and/or its safety with respect to human health and the environment.

*Study Completion Date* (FDA)

The date the final report is signed by the study director.

*Study Director* (FDA)

The individual responsible for the overall conduct of a nonclinical laboratory study.

*Study Director* (OECD)

The individual responsible for the overall conduct of the study.

*Study Initiation Date* (FDA)

The date the protocol is signed by the study director.

*Test Article* (FDA)

Any food additive, color additive, drug, biological product, electronic product, medical device for human use, or any other article subject to regulation under the act or under Sections 351 and 354–360F of the Public Health Service Act.

*Test Facility* (OECD)

The persons, premises, and operational unit(s) that are necessary for conducting the study.

*Testing Facility* (FDA)

A person who actually conducts a nonclinical laboratory study, i.e., actually uses the test article in a test system. Testing facility includes any establishment required to register under Section 510 of the act that conducts nonclinical laboratory studies and any consulting laboratory described in Section 704 of the act that conducts such studies. Testing facility encompasses only those operation units that are being or have been used to conduct nonclinical laboratory studies.

*Test Substance* (OECD)

A chemical substance or a mixture that is under investigation.

*Test Substance or Mixture* (EPA)

*EPA—FIFRSA* (Environmental Protection Agency 1984)

A substance or mixture administered or added to a test system in a study, which substance or mixture:

1. Is the subject of an application for a research or marketing permit supported by the study, or is the contemplated subject of such an application; or
2. Is an ingredient, impurity, degradation product, metabolite, or radioactive isotope of a substance described by paragraph 1. of this definition, or some other substance related to a substance described by that paragraph, which is used in the study to assist in characterizing the toxicity, metabolism, or other characteristics of a substance described by that paragraph.

*EPA—TSCA* (Environmental Protection Agency 1983)

A substance or mixture administered or added to a test system in a study, which substance or mixture is used to develop data to meet the requirements of TSCA Section 4(a) test rule and/or is developed under a TSCA Section 4 testing consent agreement or Section 5 rule or order to the extent the agreement, rule, or order references this part.

*Test System* (FDA)

Any animal, plant, microorganism, or subparts thereof to which the test or control article is administered or added for study. Test system also includes appropriate groups or components of the system not treated with the test or control articles.

*Test System* (OECD)

Any animal, plant, microbial, as well as other cellular, subcellular, chemical, or physical system, or a combination thereof used in a study.

*Validation* (ANSI)

1. The process of evaluating a system at the end of the development process to assure compliance with user requirements.
2. The process of evaluating software at the end of the software development process to ensure compliance with software requirements.

*Vehicle* (OECD)

Any agent that serves as a carrier used to mix, disperse, or solubilize the test or reference substance to facilitate the administration to the test system.

# References

Anderson, M. A. 2000. *GLP quality audit manual.* 3rd ed. Englewood, Colo.: Interpharm Press.

Carson, P. A., and N. J. Dent. 1994. *Good laboratory and clinical practices: Techniques for the quality assurance professional.* Oxford: Reed Educational and Professional Publishing Limited.

"Danish National Testing Board Act." 1973. No. 144. March 21.

Department of Health and Human Services. 1975. *Analysis of sponsor laboratories for compliance with the GLP regulations.* Washington, D.C.: US Government Printing Offices.

Hirsch, A. F. 1989. *Good laboratory practice regulations.* New York: Marcel Dekker, Inc.

"New Zealand Testing Laboratory Registration Act." 1972. No. 36. October 20.

Organisation for Economic Co-operation and Development. Revised 1997. *The OECD principles of good laboratory practice.* No.1. Paris: OECD.

US Environmental Protection Agency. 1983. *Code of Federal Regulations.* Title 40, Part 792. Toxic substances control: Good laboratory practice standards. Effective December 29. Washington, D.C.: US Government Printing Offices.

US Environmental Protection Agency. 1984. *Code of Federal Regulations,* Title 40, Part 160. Pesticide programs: Good laboratory practice standards. Effective May 2. Washington, D.C.: US Government Printing Offices.

US Food and Drug Administration. 1979. *Code of Federal Regulations.* Title 21, Part 58. Good laboratory practice for nonclinical laboratory studies. Effective June 20. Rockville, Md.: US Government Printing Offices.

# Professional Societies

The Society of Quality Assurance (SQA)
2365 Hunters Way
Charlottesville, VA 22911
Phone: 434-297-4772
Fax: 434-977-0899
www.sqa.org

Drug Information Association (DIA)
501 Office Center Drive, Suite 450
Fort Washington, PA 19034
Phone: 215-628-2288
Fax: 215-641-1229
www.diahome.org

British Institute of Regulatory Affairs (BIRA)
7 Heron Quays Marsh Wall
London E14 4JB, England, UK
Phone: 44(0)207 538 9502
Fax: 44(0)207 515 7836
www.bira.org.uk

British Association of Research Quality (BARQA)
3 Wherry Lane
Ipswich
Suffolk IP4 1LG, England, UK
Phone: 44(0)147 322 1411
Fax: 44(0)147 322 1412
www.barqa.com

American Society for Quality (ASQ)
600 N. Plankinton Ave.
PO Box 3005
Milwaukee, WI 53201
Phone: 414-272-8575
Fax: 414-272-1734
www.asq.org

For Product Safety Concerns and Information please contact our EU representative GPSR@taylorandfrancis.com
Taylor & Francis Verlag GmbH, Kaufingerstraße 24, 80331 München, Germany

www.ingramcontent.com/pod-product-compliance
Lightning Source LLC
LaVergne TN
LVHW021114300926
849160LV00009B/734

* 9 7 8 1 5 7 4 9 1 1 3 8 1 *